MY FIRST BOOK ON THE

HUMAN
BODY

A visual guide to human anatomy

Humans have more facial muscles than any other animal on this earth!

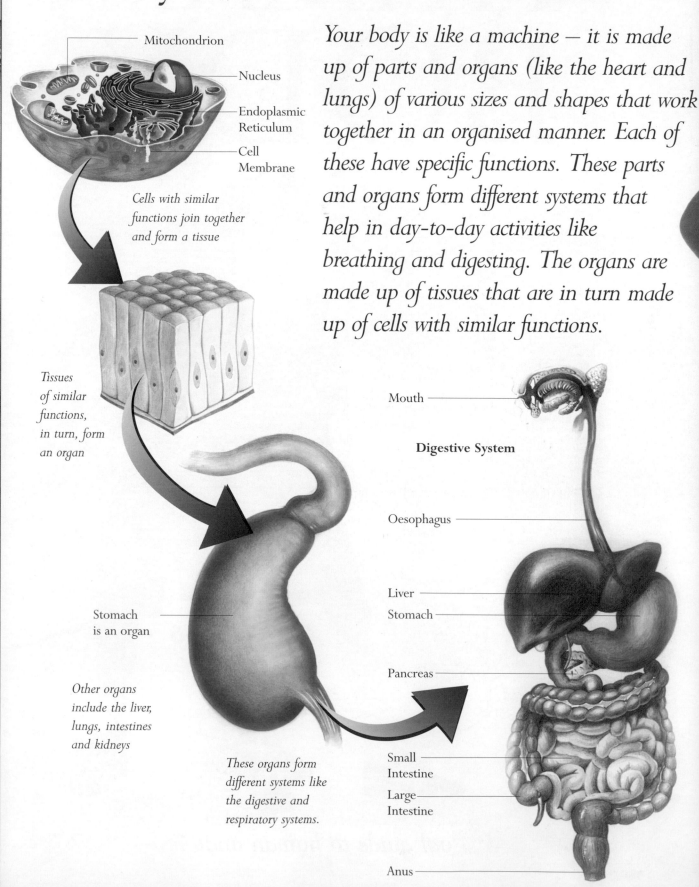

Mitochondrion

Nucleus

Endoplasmic Reticulum

Cell Membrane

Cells with similar functions join together and form a tissue

Your body is like a machine — it is made up of parts and organs (like the heart and lungs) of various sizes and shapes that work together in an organised manner. Each of these have specific functions. These parts and organs form different systems that help in day-to-day activities like breathing and digesting. The organs are made up of tissues that are in turn made up of cells with similar functions.

Tissues of similar functions, in turn, form an organ

Mouth

Digestive System

Oesophagus

Stomach
is an organ

Other organs include the liver, lungs, intestines and kidneys

Liver

Stomach

Pancreas

These organs form different systems like the digestive and respiratory systems.

Small Intestine

Large Intestine

Anus

Members of one family look like each other and in the case of identical twins, you might find it difficult to tell the two apart.

As you grow older, your body starts changing. The following chapters will tell you more about your body. Learning more about the human body and finding out how it works is a whole lot of fun!

Even though all human bodies are similar from within, we differ from each other in several ways. People differ in height, size and colour.

facts about you

- In his average lifetime, a person walks the equivalent of 5 times around the equator.
- 300 million cells die in the human body every minute.
- The average human produces 25,000 quarts of spit in a lifetime, enough to fill two swimming pools.
- Your face is made up of 53 muscles!
- Odontophobia is the fear of teeth.

The cell is the tiniest living part of your body

Cells are the building blocks of your body. They are the smallest structures that can carry out complex processes like breathing and digestion. The human body consists of around 100 trillion cells! But these cells are too small to be seen without a microscope. A cell may vary in size, shape and make-up depending upon its function. Cells can divide and multiply; that is how you grow.

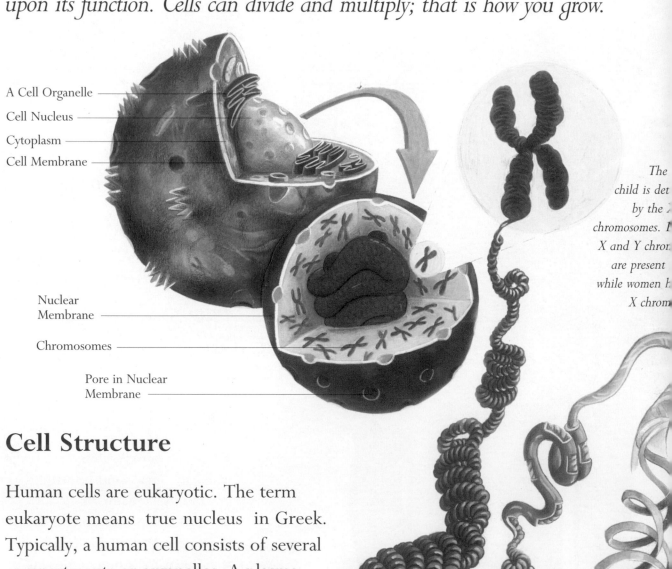

A Cell Organelle

Cell Nucleus

Cytoplasm

Cell Membrane

Nuclear Membrane

Chromosomes

Pore in Nuclear Membrane

The
child is det
by the
chromosomes.
X and Y chrom
are present
while women h
X chrom

Cell Structure

Human cells are eukaryotic. The term eukaryote means true nucleus in Greek. Typically, a human cell consists of several compartments or organelles. A plasma membrane surrounds these organelles. The organelles can be seen floating in a fluid called cytoplasm. The cells also contain a nucleus where the genetic material (DNA) is stored.

Amazing Anatomy

Is poor eyesight hereditary?

Studies have shown that poor vision is largely hereditary, i.e. it is passed on from parents to children through genes. Although it is not the only factor leading to poor vision, common problems of eyesight are said to be inherited from ones parents.

No two people in this world are alike, except for identical twins! Each one of us is unique because of our genes. But in the case of identical twins, they are alike because they share similar genes. However, you are a little like your parents because you have inherited some of their genes.

Chromosomes and DNA

Inside the nucleus lie coil-like structures called chromosomes. You have 46 chromosomes, half from your mother and half from your father. Chromosomes always come in pairs and are made up of a chemical called DNA (deoxyribonucleic acid), which looks like a spiral staircase! DNA, like the fingerprint, is unique to a person and contains all the information about that person. Sections on DNA, called genes, store traits or characters.

If the entire DNA in your body was put end to end, it would reach the sun and back over 600 times.

facts about cells

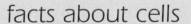

■ If all the cells in a human body were joined at the ends, they would stretch for 1000 km (650 miles) — the distance between Paris to Rome!

■ Humans have 30,000 genes.

■ As we grow older, the brain loses almost one gram a year, as its nerve cells die and cannot be replaced.

■ 50,000 of your cells will die and be replaced as you read this sentence.

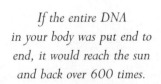

Skin is the largest organ of your body!

Your skin is a waterproof layer that protects your internal organs from infection and sunlight. It is sensitive to touch, heat and pain. It also helps to control your body temperature. Structures embedded in the skin are called skin appendages. These include hair, nails and glands. Glands produce and secrete substances needed by other body parts. The sweat glands carry the sweat to the surface of the skin where it is let out through tiny holes called sweat pores.

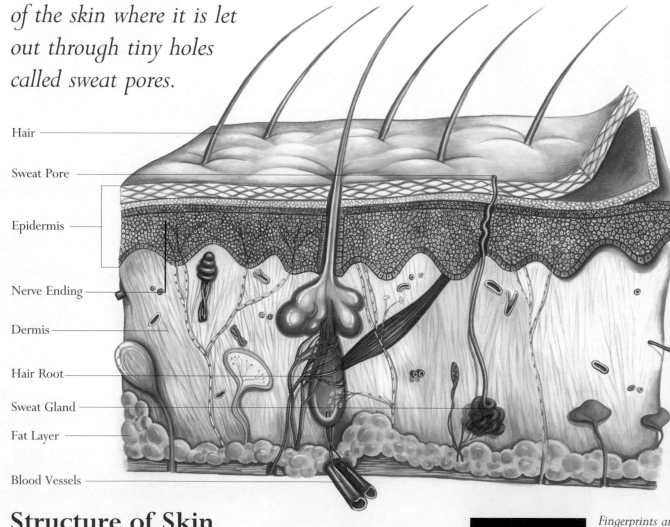

Hair

Sweat Pore

Epidermis

Nerve Ending

Dermis

Hair Root

Sweat Gland

Fat Layer

Blood Vessels

Structure of Skin

Your skin is made up of two layers of tissue. The outer layer is called the epidermis and the layer below is called the dermis. A thick coating of a tough, fibrous protein called keratin covers the epidermis. The dermis contains tiny blood vessels, sweat glands, nerve endings and the roots of tiny hairs. Under the dermis is a layer of fat, which keeps you warm.

Fingerprints are thin ridges of skin on the fingertips. These ridges could be in the form of loops or lines. No two people have the same fingerprints.

Amazing Anatomy

Why are some people dark and some fair?

Your skin gets its colour from a pigment called melanin, which is produced by a network of special cells called melanocytes. These cells also help the skin from harmful sunrays. People are darker in the warmer regions because their skin produces more melanin to protect them from harsh sunlight.

Hair and Nails

Like skin, your hair and nails are also made of keratin. Each hair is born in a tiny pit called a follicle. The roots of your hair are alive but the part above the skin is made up of dead cells. That is why you don t feel pain while cutting your hair! Hair gets its colour from melanin.

Each nail grows about a millimetre every 10 days. The new nail forms behind the cuticle, which is under the skin and pushes the older nail outside. Like your hair, nails too are made up of dead cells.

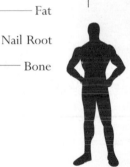

Nail

Skin

Cuticle

Fat

Nail Root

Bone

Some people have freckles. These are small patches of darker skin made by extra melanin. Exposure to sunshine increases the amount of melanin in your skin and the darkness of freckles.

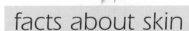

facts about skin

- Each person loses about 80 hairs every day.
- The entire length of all the eyelashes shed by a human in their life is over 30 m (98 feet).
- The skin of a normal adult male weighs 4.5-5 kg (10-11 pounds) and measures 2 m² (22 square feet)
- Each square inch (2.5 cm) of human skin consists of 6 m (20 feet) of blood vessels.

The number of bones in your neck is the same as a giraffe s!

Your body gets its shape from a framework of bones called the skeleton. Without the skeleton, you would be shapeless! Its function is also to support and protect the delicate organs inside the body. For example, the skull protects the brain and the ribs protect the heart and lungs; while the teeth provide support to the facial muscles and help form a smile! On its own, each bone is hard and rigid but together the skeleton is flexible and helps you move.

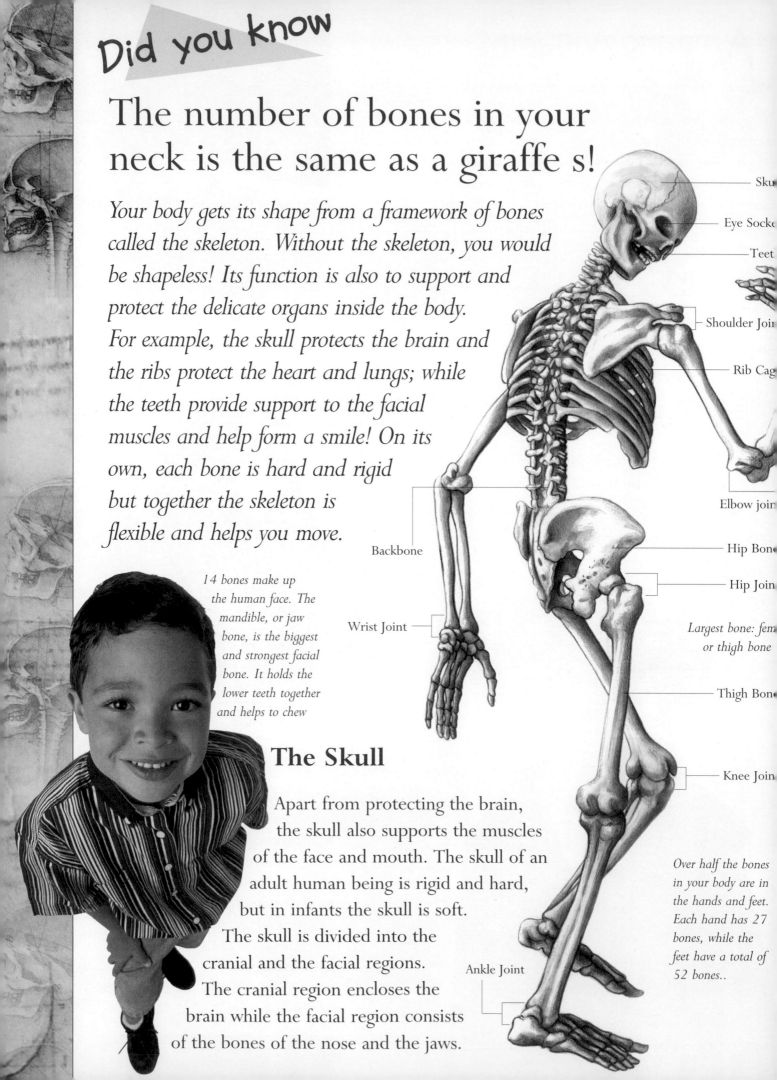

14 bones make up the human face. The mandible, or jaw bone, is the biggest and strongest facial bone. It holds the lower teeth together and helps to chew

Sku
Eye Socke
Teet
Shoulder Joir
Rib Cag
Elbow join
Hip Bon
Hip Join

Largest bone: fem or thigh bone

Thigh Bon

Backbone

Wrist Joint

Knee Join

The Skull

Apart from protecting the brain, the skull also supports the muscles of the face and mouth. The skull of an adult human being is rigid and hard, but in infants the skull is soft. The skull is divided into the cranial and the facial regions. The cranial region encloses the brain while the facial region consists of the bones of the nose and the jaws.

Ankle Joint

Over half the bones in your body are in the hands and feet. Each hand has 27 bones, while the feet have a total of 52 bones..

Amazing Anatomy

What is a cavity?

If you allow food to remain stuck in your teeth for a long time, the bacteria in your mouth feed on it. In the process, the bacteria destroy the enamel, causing a hole in your tooth. This is called a cavity. Brushing and flossing your teeth regularly can prevent cavities.

You must brush your teeth twice a day for at least two minutes. You should also change your toothbrush every three months.

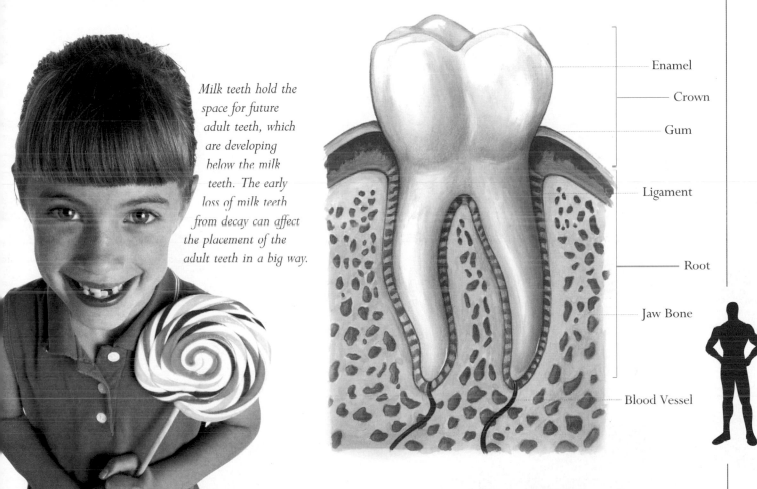

Milk teeth hold the space for future adult teeth, which are developing below the milk teeth. The early loss of milk teeth from decay can affect the placement of the adult teeth in a big way.

Enamel
Crown
Gum
Ligament
Root
Jaw Bone
Blood Vessel

Teeth

Every tooth comprises the root that anchors it in the jawbone and the visible part called the crown. Teeth are covered with enamel, which is the hardest substance in the human body. A baby has about 20 teeth by the time he is 2 years old. These are called milk teeth, and start to fall out at about the age of 6. The milk teeth are replaced by a set of permanent teeth.

facts about skeletons

■ Your skull is made up of 29 different bones.
■ Babies are born with about 300 bones, some of which fuse together as they grow.
■ Jaw muscles can provide about 200 pounds of force to bring the teeth at the back together for chewing.
■ The feet account for one quarter of all the bones in the human body.

Most of your body s calcium is contained in your bones and teeth!

Each bone in your body is hard and rigid outside and lighter and softer inside. The bones are hard because they are made up of calcium and phosphorus. Many bones are hollow making them light. The softer inner tissue found in the centre is called the marrow. The marrow produces red and white blood cells and stores fat. Most people have 206 bones but some may have extra bones in their thumbs or big toes.

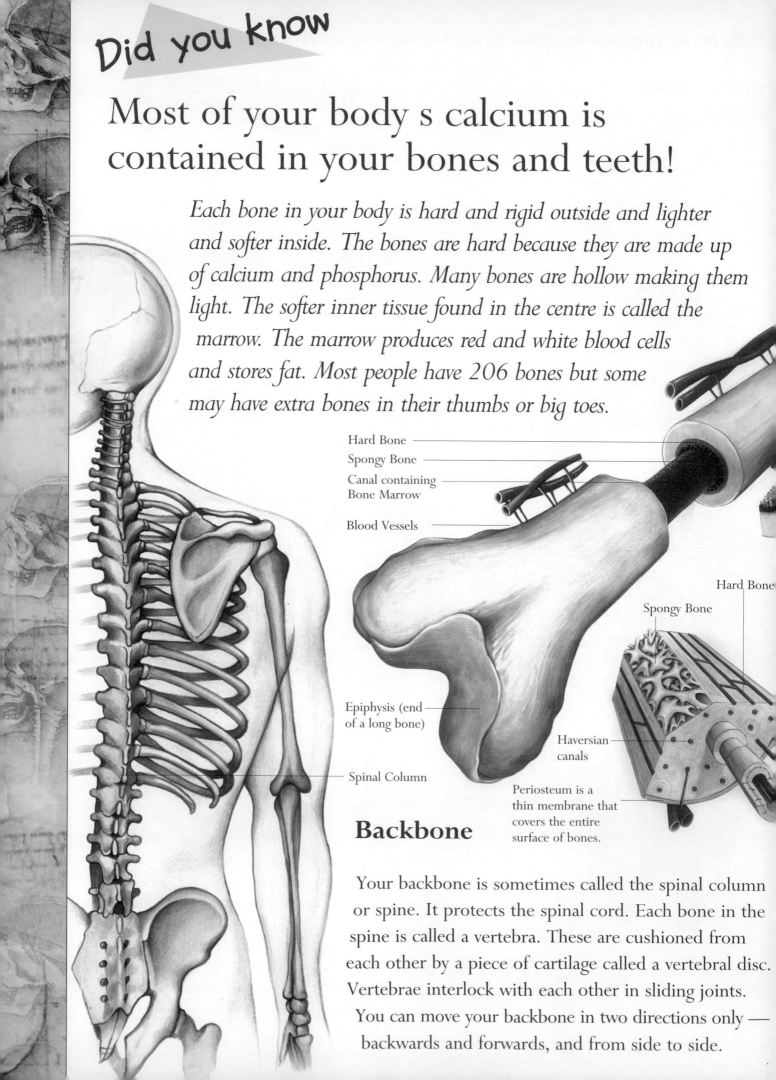

Hard Bone

Spongy Bone

Canal containing Bone Marrow

Blood Vessels

Hard Bone

Spongy Bone

Epiphysis (end of a long bone)

Haversian canals

Spinal Column

Periosteum is a thin membrane that covers the entire surface of bones.

Backbone

Your backbone is sometimes called the spinal column or spine. It protects the spinal cord. Each bone in the spine is called a vertebra. These are cushioned from each other by a piece of cartilage called a vertebral disc. Vertebrae interlock with each other in sliding joints. You can move your backbone in two directions only — backwards and forwards, and from side to side.

Amazing Anatomy

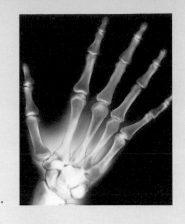

What is a fracture?

A fracture is a break or a crack in the bone. The bone must be repaired before you can use it again. A doctor will firmly fasten your bone in a plaster so that the ends do not move and it sets easily. A few days after the fracture, a fibrous network is formed by certain cells in the damaged area. Then cartilage is produced which is in turn converted to bone by repairing cells called osteoblasts.

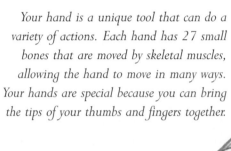

Your hand is a unique tool that can do a variety of actions. Each hand has 27 small bones that are moved by skeletal muscles, allowing the hand to move in many ways. Your hands are special because you can bring the tips of your thumbs and fingers together.

Joints

A joint is the point where two or more bones meet. Bands of tissues called ligaments hold most joints together. Your body has several types of joints such as the ball-and-socket joint in your shoulder that helps your arms move, or the hinge joints which are the joints between the bones in your knees and elbows. These bones can only move to and fro like doors.

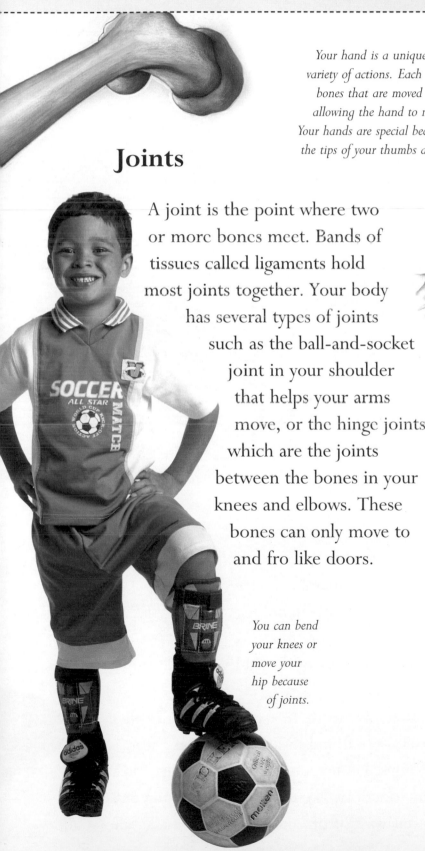

You can bend your knees or move your hip because of joints.

facts about bones

■ The stapes or stirrup in the middle ear is the smallest bone. It is about the size of a grain of rice. It is only 0.18 cm (0.07 inches) long.
■ The femur or thigh bone which is the longest bone in the body grows up to 50cm (20 inches).
■ The thyroid cartilage is more commonly known as Adam s apple.
■ Babies are born without kneecaps. They do not form until the child reaches 2-6 years of age.

Muscles help you blink!

You can hop, skip, blink and even breathe because of your muscles. Your body has more than 600 muscles. These are mainly of three types: skeletal muscle, smooth muscle and cardiac muscle.

Muscles work in opposing pairs such as the biceps and triceps in your arm. When you raise your arm, the biceps contracts and shortens while the triceps expands.

The opposite happens when you lower your arm. Muscles have cells that help them expand and contract. These cells use chemical energy from the food you eat to do this.

A bodybuilder does special exercises to strengthen his muscles. Lifting weights make the muscles in his arms hard and strong by improving the blood circulation. He also eats the right food to get more energy.

Skeletal and Smooth Muscles

Skeletal Muscle Fi

Skeletal muscles are those that you can move by choice (voluntary muscles). They are sausage-like fibres surrounded by membranes. Most of them are connected to the skeleton by tissues called tendons. The skeletal muscles help move the various bones and cartilages of the skeleton.

Nucleus

Smooth Muscle Fibre

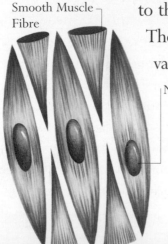

Nucleus

Smooth muscles are those that work automatically. These muscles are found in organs such as the stomach, lungs, kidneys and skin. They help you in day-to-day functions like digestion, breathing and throwing out waste from the body.

Amazing Anatomy

Can you hurt your muscles?

Yes, you can pull a muscle just like you can tear a ligament or break a bone. Muscles can repair themselves with care and proper nutrition.

Most muscles are linked to bones by strong cords called tendons. A ballerina can make swift and graceful movements using her foot muscles.

Cardiac Muscle Fibre

Nucleus

Cardiac Muscle

This is a special tissue found only in your heart. Though the heart also contains smooth muscles, its functions are mostly performed by the cardiac muscle. Unlike other muscles in the body the cardiac muscle never gets tired. It constantly works to pump blood in and out of the heart.

Tendons

It takes twice the amount of muscles to frown than it takes to smile.

facts about muscles

■ You have over 30 facial muscles that help create looks of surprise, happiness, sadness and anger.
■ Eye muscles are the busiest muscles in the body. Scientists estimate they may move more than 100,000 times a day!
■ You use more than 200 different muscles when you walk!
■ The muscle in your buttocks is the largest muscle in your body. It is called the gluteus maximus!

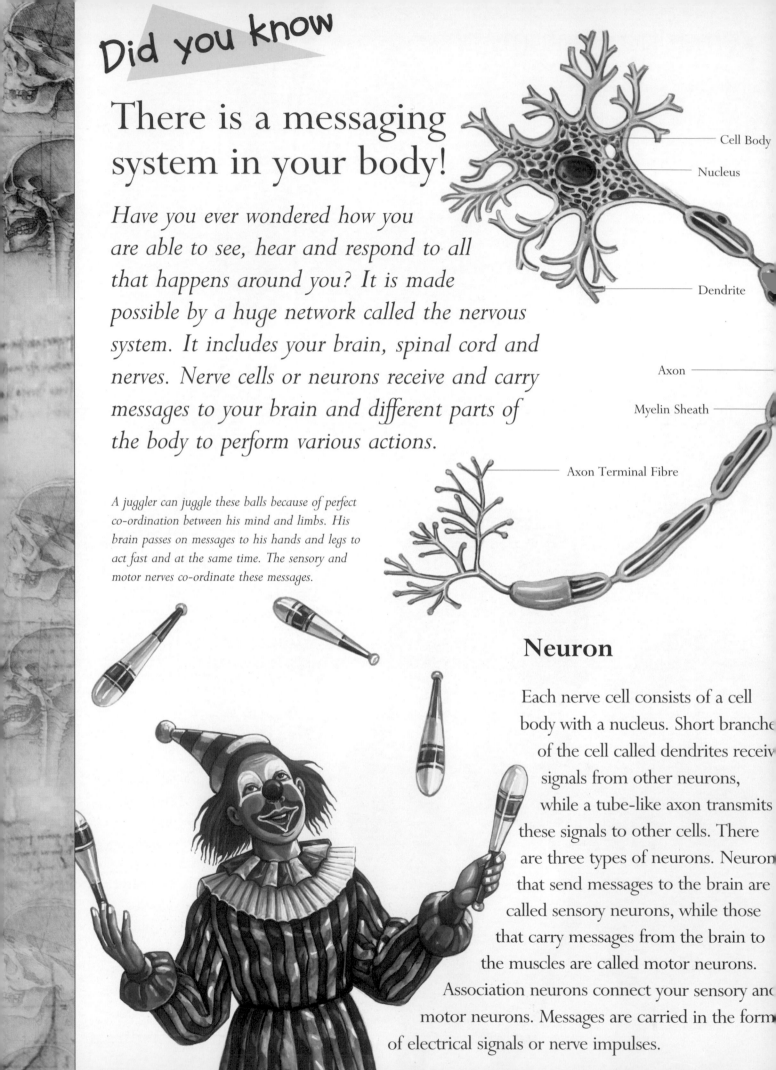

There is a messaging system in your body!

Have you ever wondered how you are able to see, hear and respond to all that happens around you? It is made possible by a huge network called the nervous system. It includes your brain, spinal cord and nerves. Nerve cells or neurons receive and carry messages to your brain and different parts of the body to perform various actions.

A juggler can juggle these balls because of perfect co-ordination between his mind and limbs. His brain passes on messages to his hands and legs to act fast and at the same time. The sensory and motor nerves co-ordinate these messages.

Cell Body

Nucleus

Dendrite

Axon

Myelin Sheath

Axon Terminal Fibre

Neuron

Each nerve cell consists of a cell body with a nucleus. Short branches of the cell called dendrites receive signals from other neurons, while a tube-like axon transmits these signals to other cells. There are three types of neurons. Neurons that send messages to the brain are called sensory neurons, while those that carry messages from the brain to the muscles are called motor neurons. Association neurons connect your sensory and motor neurons. Messages are carried in the form of electrical signals or nerve impulses.

Amazing Anatomy

Why do you feel pain?

You feel pain when your free nerve endings are damaged. Free nerve-endings in your skin may pick up a sensation or stimulus and send the message up the spinal cord to the brain. The points on the skin that respond to stimuli with a sense of pain are called pain spots . These spots are not evenly distributed in the body. Some places might have fewer pain spots than others.

Spinal Cord

The spinal cord is a bundle of nerves and is around 40 cm long. It is connected to the brain and extends down the backbone. Messages are conveyed to and from the brain through the spinal cord. In certain cases the spinal cord too conveys messages. For example, if you touch something hot, your hand moves away automatically. You blink when there is too much light. These are reflex actions. Most of these reflexes do not travel to your brain. Instead your spinal cord sends out the necessary messages. That is why they take place so quickly!

There are about 3,000,000 spots on the skin where pain can be felt!

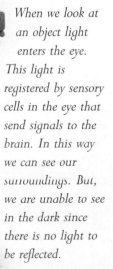

When we look at an object light enters the eye. This light is registered by sensory cells in the eye that send signals to the brain. In this way we can see our surroundings. But, we are unable to see in the dark since there is no light to be reflected.

facts about nerves

■ Nerve impulses to and from the brain travel as fast as 274 km (170 miles) per hour.

■ The most sensitive cluster of nerves is at the base of the spine.

■ The longest nerve in the body is the sciatic nerve. It stretches from the base of the spine to the knee.

■ Some people cannot feel pain — this disease is called Syringomyelia .

■ The length of all your nerves put together would be 72 km (45 miles)!

An adult eyeball is about the size of a golf ball, most of it hidden inside your head!

There are five sense organs in the human body. They are the ear, nose, eyes, skin and tongue. Your eyes help you see the world around while your ears help in hearing. However, the actual processes are carried out by the brain. The eyes convert the light waves and the ear converts sound waves into nerve impulses before sending them to the brain.

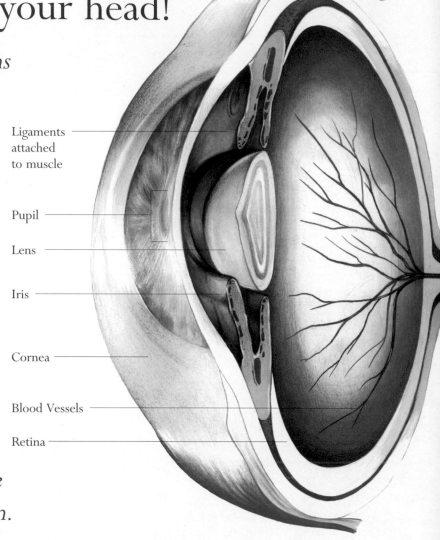

Ligaments attached to muscle

Pupil

Lens

Iris

Cornea

Blood Vessels

Retina

Eyes

You see an object when the light bouncing off the object enters your eyes. Light rays enter the eye through a small opening called the pupil. They are then focussed by the lens to form an image on the retina.

Cells sensitive to light trigger nerve signals that travel to the brain. The picture formed on the retina is upside down, but the brain turns it the right way up. Light cells are of two kinds — rods and cones. Rods help to see shades of grey and in the night. Cones on the other hand detect colours.

Amazing Anatomy

Why can not you see in the dark?

When we look at an object light enters the eye. This light is registered by sensory cells in the eye that send signals to the brain. In this way we can see our surroundings. But we are unable to see in the dark since there is no light to be reflected.

Ears

Only a part of your ear is seen outside, the rest is inside. The outer ear has a flap called the pinna and the ear canal. The middle ear consists of three tiny bones - the hammer, the anvil and the stirrup. The eardrum separates the outer and middle ear. The inner ear contains the spiral shaped cochlea, the vestibule and three fluid-filled semicircular canals that help maintain your balance.

The earflap directs sound waves into the ear canal. The waves then bounce off the eardrum making it vibrate. The three tiny bones magnify the sound vibrations and send them to the cochlea. Millions of tiny hairs then convert these vibrations into electric signals that are sent to the brain.

The loudness of sound is measured in decibels (dB). Sounds above 130dB can cause damage to the ear and may cause deafness.

facts about senses

■ Women blink nearly twice as much as men.

■ Our eyes are always the same size from birth, but our nose and ears never stop growing.

■ Glands in the ear produce ear wax which protects the eardrum from dirt and dust. Its unpleasant smell stops insects from entering the ear!

■ Male eyes are about 0.5 mm bigger than female eyes.

No two persons have the same tongue print!

You already know about your eyes and ears. Now let's read about the other sense organs — namely skin, nose and tongue. Although your sense of smell is stronger than your sense of taste, these two actions are closely linked to each other. For example, during a bad cold even the most delicious food tastes bland because you are unable to smell the aroma of the food!

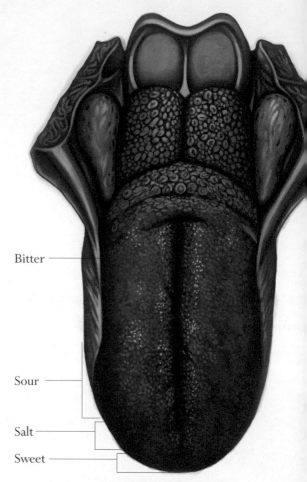

Bitter

Sour

Salt

Sweet

Nose and Tongue

Your nose has special nerve endings at the back that are called olfactory receptors. When a smell dissolves in the mucus inside your nose, the hairs on the receptors take it in. This excites the receptors, which then send a message to the brain through a nerve called the olfactory nerve.

Your tongue is covered with taste buds. Each of these taste buds contains several receptors with tiny hairs that feel the food. Every taste bud can make out different tastes such as sweet, sour, bitter and salty. Your tongue also plays a vital role in speech by helping to form words.

Amazing Anatomy

What is saliva?

Saliva is a watery liquid secreted by glands present in the mouth. It moistens the food and starts the process of digestion even before the food is swallowed. Taste buds only work on the chemicals in food after it has been dissolved by the saliva.

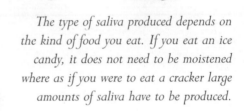

Touch

Your sense of touch includes responses to pain, pressure and temperature. You can tell the difference between hot and cold and wet and dry. This is possible because of the receptors in your skin. These receptors are actually ends of nerves, which respond to your environment. They convert touch into tiny electrical nerve signals and send them to your spinal cord and brain. There are different receptors for different sensations like pain, pressure and temperature.

The type of saliva produced depends on the kind of food you eat. If you eat an ice candy, it does not need to be moistened where as if you were to eat a cracker large amounts of saliva have to be produced.

The dermis is filled with many tiny nerve endings which give you information about the things with which your body comes in contact. They do this by carrying the information to the spinal cord, which sends messages to the brain where the feeling is registered.

facts about senses

- The human tongue is about 10 cm (4 inches) long.
- Fish have taste buds all over their body — so they can taste even with their fins!
- The back is one of the least sensitive areas of the body — least sensitive to touch!
- When you sneeze, all your bodily functions — even your heart — stops!

Human brains are more complex than those of any other animal!

Your brain is the most important organ of the nervous system. It monitors and controls body processes like breathing, heart rate and digestion. It is also the seat of learning, thinking and even feelings. Your brain is protected by the skull. The brain can be broadly divided into the brain stem, the cerebellum, the cerebrum and the diencephalon.

Longitudinal Fissure

Left Cerebral Hemisphere

Right Cerebral Hemisphere

Cerebrum and Cerbellum

The front portion of the cerebrum is involved in speech, thought, feelings and logical activities like maths. The part behind this is responsible for understanding touch, while the section on both sides of the cerebrum help in hearing and storing memory. The cerebrum is divided into two hemispheres. The right half is responsible for artistic and creative activities such as music or painting. The cerebellum controls the muscles and helps in keeping the balance of the body.

Amazing Anatomy

Does the brain work while sleeping?

When you sleep, your muscles relax and your body rests. But your brain is at work when you sleep. Scientists believe that it sorts out information and solves problems. In fact some think that dreams are your brain's way of making sense of what happened during the day. It is said that dreams are a clue to what you're worried or anxious about.

The Brain Stem and Diencephalon

The brain stem controls unconscious or involuntary processes like digestion, breathing and heartbeat. The diencephalon is found on top of the brain stem and consists of the thalamus and the hypothalamus. The thalamus receives information from the sense organs and sends it to the appropriate part of the brain. The hypothalamus maintains the body conditions by controlling thirst, hunger and body temperature.

You should always wear a helmet while skating, biking or riding to prevent any serious head injuries.

The part of your brain responsible for processing memory is the hippocampus. The hippocampus is a brain structure which lies under the medial temporal lobe found on each side of the brain

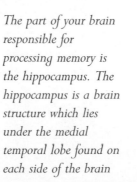

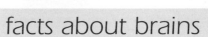
facts about brains

■ The human brain comprises about 85% water.
■ A newborn baby's brain grows almost 3 times in size in the first year!
■ The left side of your brain controls the right side of your body; and the right side of your brain controls the left side of your body.
■ The average person moves about 20 to 40 times a night, but only about 30 seconds in each hour.

The heart is a fist-sized muscle in your body!

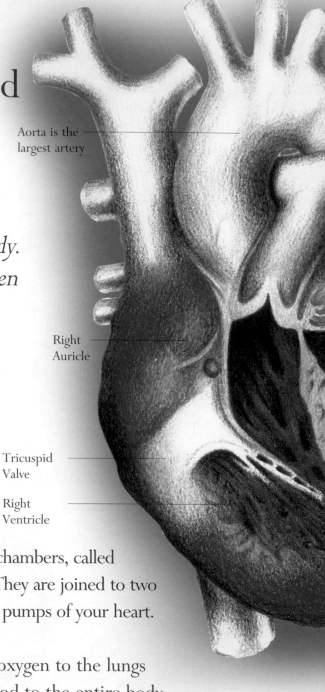

The heart is to your body what an engine is to a car! Your heart pumps oxygen-rich blood throughout the body. It also sends the blood with less oxygen to your lungs. The heart is made up of a special kind of muscle called the cardiac muscle and never gets tired. It hardly rests for a second between beats and constantly pumps blood.

Aorta is the largest artery

Right Auricle

Tricuspid Valve

Right Ventricle

Structure of the Heart

Your heart has four chambers. The upper two chambers, called atria or auricles, receive blood from the veins. They are joined to two lower chambers called ventricles. These are the pumps of your heart.

The right ventricle pumps blood that needs oxygen to the lungs and the left ventricle pumps oxygen-rich blood to the entire body. The muscular walls of the left ventricle are thicker than the walls of the right ventricle making it easier to feel your heart beat on the left side of your chest.

A tiny group of cells called the sinoatrial node co-ordinates the contracting of the cardiac muscles in the various parts of the heart. This sinoatrial node sends an electrical signal around the heart just before every beat. An electrocardiograph (ECG) machine can record these impulses. Doctors can read your heartbeat using this machine.

Amazing Anatomy

What is a heartbeat?

There are one-way valves between the chambers in your heart that keep the blood flowing in the right direction. As blood flows through a valve from one chamber to another, the valve snaps shut preventing the blood from flowing backwards. When the valves snap shut, they make a thumping sound - this is what you call a heart-beat!

Left
Auricle

Systole and Diastole

There are two stages in a heart beat — the systole and the diastole. Systole is the condition when the heart pumps blood into the arteries. In this process the walls of the ventricles contract. This raises the blood pressure inside the ventricles. As a result the blood is forced into the arteries.

During diastole the muscles of the ventricles relax. The decrease in the pressure causes the one-way valve in the heart to open. Meanwhile the atrium contracts forcing blood through the opening into the ventricles.

Your heart rate increases when you exercise because your muscles need extra oxygen during the work out.

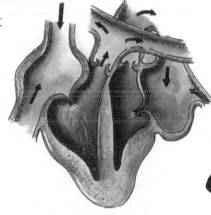

This is how your heart pumps blood into other parts of your body!

Left
Ventricle

Left Auricle
Right Auricle

Left Ventricle
Right Ventricle

facts about hearts

■ In one day your heart transports all your blood around your body about 1000 times.

■ The human heart creates enough pressure to squirt blood 9 m (30 feet) high!

■ In an average lifetime, the human heart will pump 169 million litres of blood.

■ The left lung is smaller than the right lung to make room for the heart.

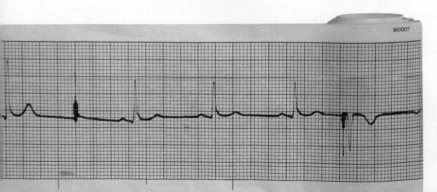

Blood is red in colour but some of its cells are white!

The cells in your body need a steady supply of oxygen and nutrients to survive. At the same time, carbon dioxide and other waste materials produced by these cells have to be removed. The blood plays a major role in these processes. Your heart forces blood through a network of tube-like blood vessels running throughout the body. Your bloodstream absorbs and transports oxygen, nutrients and water to all the body cells. It also picks up the body's waste products and transports them to the kidneys and lungs, where the waste products are released from the body.

Kidney

Blood Composition

Blood mostly consists of a yellow fluid called plasma. The rest is made up of red blood cells (erythrocytes), white blood cells (leukocytes) and cell fragments called platelets (thrombocytes). Proteins, hormones, salt and water are dissolved in the plasma. Red blood cells are composed of a protein called haemoglobin, which absorbs oxygen as blood passes through the lungs and releases it to the rest of the body. Blood cells are produced in the bone marrow, specifically in the spine, ribs, hips, skull and the breastbone (sternum).

The white blood cells fight disease and infection. Different types of white cells live for different life spans.

Bone marrow

Red blood cells

Amazing Anatomy

What are the different types of blood groups?

There are four main types or groups of blood: A, B, AB and O. Blood cells and some of the special proteins blood contains can be replaced or supplemented by giving a person blood from someone else. This process is called transfusion . But transfusion can only take place between similar blood groups.

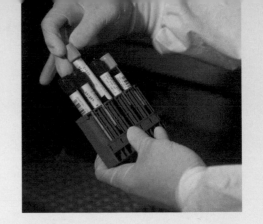

Movement of Blood

Blood travels to all parts of the body through a network of blood vessels. Arteries are large, strong and thick-walled blood vessels that carry oxygen-rich blood from the heart to the rest of the body. Veins are smaller and thinner blood vessels that carry blood deficient in oxygen back to the heart. Tiny branches of blood vessels called capillaries connect arteries with veins in body tissues. Most capillaries are so narrow that only one red blood cell can fit through them at a time!

Blood pressure is the pressure on the walls of arteries applied by the blood it carries. Blood pressure is low when we sleep and rises during exercise. Rushing blood makes some people look redder during rigorous activity

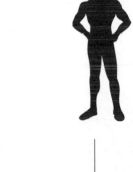

Artery

White Blood Cells

Platelets

Red Blood Cells

facts about blood

■ On an average, red blood cells live up to four months after being released by the bone marrow.
■ Platelets are the smallest cells in the blood and cannot live longer than ten days.
■ Every square inch (2.5 cm) of human skin consists of 6 m (20 feet) of blood vessels.

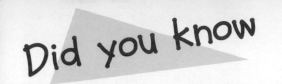

You can survive with just one fully working lung!

Your body needs oxygen to survive. You get this oxygen from the air you breathe. This involves a very complicated process in which the lungs play a vital role. Your lungs are large spongy organs that occupy most of the chest cavity. When blood passes through the lungs, oxygen is absorbed by the haemoglobin and circulated to all parts of your body.

Nose | Nasal Cavity | Larynx

Trachea or Wind pipe

Bronchi

Right Lung

Left Lung

Heart

Rib

Inside the lungs

The air that you breathe through the nose and mouth enters the trachea or windpipe. The trachea in turn divides into two tubes called bronchi that further branch out into smaller tubes inside the lungs. These tubes end in tiny bubbles or sacs called alveoli. The alveoli pass on the oxygen from freshly inhaled air to your bloodstream. They also exchange it for waste products such as carbon dioxide that has been produced in the body.

Amazing Anatomy

Why does our breath form a mist in winters?

The air from our lungs is warm and moist. When it mixes with the cold winter air, it cools, forming tiny droplets. You can see these droplets in the form of a mist or fog.

Air pollution is the contamination of air by the discharge of harmful substances like fumes from motor vehicles. It can cause health problems like burning eyes and nose, itchy, irritated throat and breathing problems. Some chemicals found in polluted air can even cause cancer, birth defects, and brain and nerve damage! Air pollution also damages the environment and property.

How you breath

You take air in about 20 times in a minute. When you take air in (inhale), the muscles around your rib cage thrust your ribs out and lower the dome of your diaphragm, increasing the space in your chest. This, in turn, decreases the air pressure inside the lungs and air flows into your lungs. When the muscles relax, the ribs and the diaphragm return to their original position, decreasing the chest cavity. This causes your lungs to push out the extra air inside. This is called exhalation.

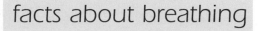

Humans cannot breathe in the oxygen dissolved in water. Scuba divers need oxygen cylinders to breathe under water.

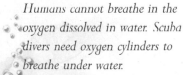

facts about breathing

- You breathe over six litres (1.32 gallons) of air every minute.
- If the inner surfaces of the lungs were laid out flat, their total area would be 180 m² (1, 938 sq ft.)
- Your lungs contain around 800 million alveoli!
- During a 24-hour period, the average human will breathe approximately 23,040 times.

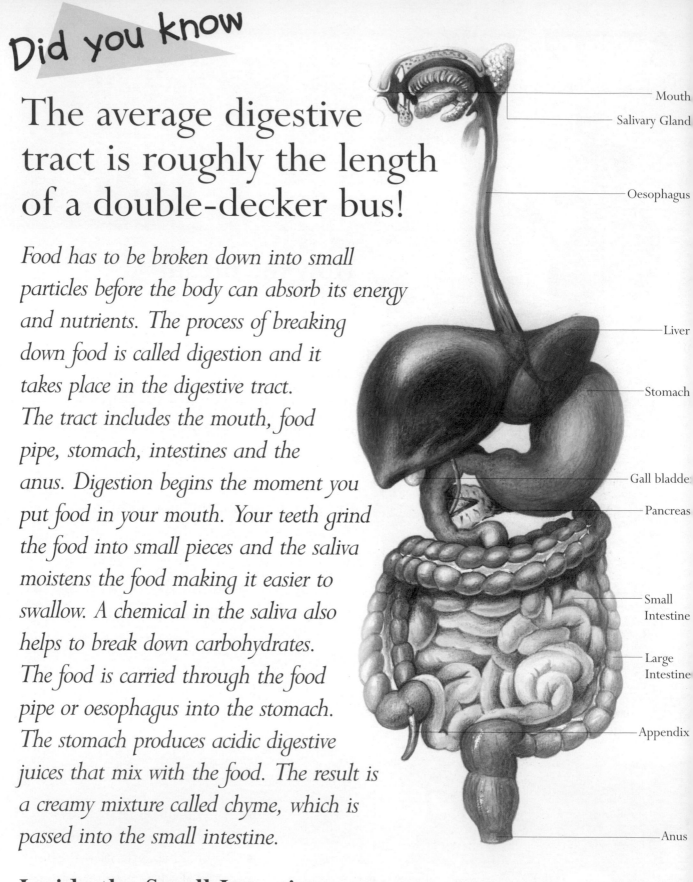

The average digestive tract is roughly the length of a double-decker bus!

Food has to be broken down into small particles before the body can absorb its energy and nutrients. The process of breaking down food is called digestion and it takes place in the digestive tract. The tract includes the mouth, food pipe, stomach, intestines and the anus. Digestion begins the moment you put food in your mouth. Your teeth grind the food into small pieces and the saliva moistens the food making it easier to swallow. A chemical in the saliva also helps to break down carbohydrates. The food is carried through the food pipe or oesophagus into the stomach. The stomach produces acidic digestive juices that mix with the food. The result is a creamy mixture called chyme, which is passed into the small intestine.

Mouth

Salivary Gland

Oesophagus

Liver

Stomach

Gall bladder

Pancreas

Small Intestine

Large Intestine

Appendix

Anus

Inside the Small Intestine

A large part of digestion, including the absorption of nutrients, happens in the small intestine. Bile from the liver enters the small intestine and breaks down fat. Pancreatic juice breaks down sugars and starch and eventually the enzymes in the small intestine complete the process. The nutrients are then absorbed through the walls of the small intestine and carried into your bloodstream.

Amazing Anatomy

How is food converted into energy?

Food undergoes the process of digestion from the moment it is put in your mouth. The complex food molecules like carbohydrates are broken down into simpler molecules like glucose, which are then absorbed into the blood. The blood carries these food nutrients to all parts of your body giving them energy for their various functions.

Inside the Large Intestine

The large intestine begins at the colon where some of the remaining nutrients like water and salts are absorbed. The rest of the waste matter is then stored in the rectum until it is excreted from the body.

Bread - source of Carbohydrate

Cheese - source of Fat

Mushroom - source of Iron

Grapes - source of Vitamin C

Milk - source of Calcium

Cauliflower - source of Minerals

Carrot - source of Vitamin A

Fish - source of Iron and Zinc

Egg - source of Protein

Eating Healthily

A healthy diet would consist of lots of vegetables and fruits. Moderate helpings of meat and fish and very little fat will help you to keep fit. Vegetables and fruits are excellent sources of vitamins, minerals and fibre. Milk is a complete food and will provide you with all the essential nutrients. Meat, fish and eggs are good sources of iron and zinc.

facts about eating

■ Your stomach needs to produce a new layer of mucus every two weeks or it would digest itself!

■ If uncoiled, the small intestine would be about 6 m (20 feet) long.

■ The human body has enough fat to produce 7 bars of soap!

■ The digestive tract is around 9 m (30 feet) long.

■ It takes 3 hours for food to move through the intestine.

You can survive with one kidney. If one stops working the other enlarges to handle all the work!

Your body produces waste that cannot be used. So what does it do with this waste? These products are passed into your kidneys where they are expelled from the body along with excess water in the form of urine. The semi-solid waste is stored in the rectum and later expelled from the body through the anus.

Sweat is also known as perspiration. It is made of water, tiny amounts of ammonia, urea, salts and sugar. Sweat leaves your skin through tiny holes called pores. When it evaporates off your skin, your body cools down. Sweating helps your body to remain cool during hot summers.

Kidney

Renal Artery

Renal Vein

Kidneys

Kidneys are bean-shaped organs containing millions of small tube-like structures called nephrons. These nephrons filter out excess water, salt and other wastes from the blood, and reabsorb some of them. The rest is used to produce urine which is then drained into tubes called ureters leading to the urinary bladder. When your bladder becomes too full, you feel the need to pass urine. The bladder can expand to hold almost a pint of urine. It also closes openings into the ureters, so that urine cannot flow back into the kidneys. The tube through which the urine flows out of the body is called the urethra.

Amazing Anatomy

Why do babies wet the bed?

When the urinary bladder is full, the nerves in the bladder send signals to the brain, making you want to go to the bathroom. Once in the bathroom the brain sends signals and the bladder squeezes the urine out. A group of muscles called the pelvic floor muscles keep the urine in until we reach the bathroom. In babies, however, this co-ordination between the brain and the bladder is not developed.

Burps are the sounds that you make when extra gases escape your stomach very quickly, to pass through the oesophagus and out of your mouth.

Large Intestine, Rectum and Anus

All the leftover waste that cannot be digested enters the large intestine. Here, water is reabsorbed causing the waste to harden. This is then stored in a part of the large intestine called the rectum in the form of a stool. The stool is pushed into the anus and out by the contraction of the muscles in the rectum. Bacteria present in the intestines feed on the waste producing two chemicals called indole and skatole. These chemicals give the stool its bad smell.

facts about excretion

- Each kidney contains about 1.3 million nephrons.
- Around 180 litres of blood is filtered by the kidneys to produce 1.5 litres of urine every day!
- You belch or fart at least 10 — 15 times a day!
- Escaping gases causes farts or belches. These can travel through your body in about 30 to 45 minutes. Burps are even faster!

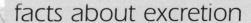

URINARY SYSTEM

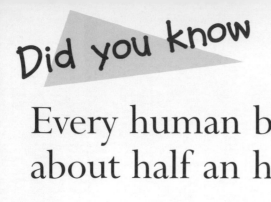

Every human being has spent about half an hour as a single cell!

All living beings, whether plants or animals, produc[e] young ones. Humans, too, produce their y[oung] ones. This process is called reproduction. A new human life is born when a father's sperm (male reproductive cell) combines with an ovum (female egg) from the mother. The new cell that is formed divides into many cells to form a baby.

Embryo

After fertilisation, the egg divides into a morula or a ball of cells. Almost a week later they form a hollow ball called the blastocyst. Soon the blastocyst develops to form a placenta, umbilical cord and the developing baby or embryo inside the mother s womb. The baby gets its food and oxygen from the mother s bloodstream through the placenta that is attached to the baby s umbilical cord. The embryo floats in a fluid called the amniotic fluid that protects it from bumps.

Foetus

Umbilical cord

Amniotic fluid

Wall of Uterus

Amazing Anatomy

Why do babies cry?

Crying is a baby's way of communicating. A normal healthy baby may cry between one and three hours per day. By the time the baby is 10 to 14 days old, parents start to make out the different types of cries — hunger, pain, tiredness or even boredom!

About 10 percent of babies cry excessively — more than 3 hours a day!

Foetus

By the eighth week the embryo is 2.5cm long. At this point it is called a foetus. It takes twelve weeks for the embryo to develop all the internal organs. During the last six months the baby starts to develop fingernails and hair. It starts to move, kick and suck its thumb. It can even tell light from dark! The baby is ready to be born after about 40 weeks or nine months.

When the baby is to be born, the mother experiences labour pains as the muscles of her uterus contract and the cervix dilates to push the baby out.

A baby's skull is really delicate; it becomes strong over a period of time.

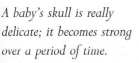

At six months babies can usually sit themselves up and even make certain sounds. Their first teeth also start to grow now.

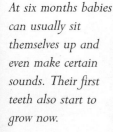

facts about babies

- The ovum or egg is the largest cell in the human body. It is about 2 inches in diameter.
- The smallest cell in the human body is the male sperm. It takes about 175,000 sperm cells to weigh as much as a single egg cell.
- A baby's head measures about one quarter of its body length. When it reaches adulthood, the head is about one eighth of the length of its body!

You are taller when you wake up than during the day!

During the day the discs of the spine get compressed (squeezed) due to gravity, making you just a tiny bit shorter.

You begin life as a baby and slowly grow up over the years. When you are about six or seven years old, your milk teeth are replaced by a set of permanent teeth. You start to grow taller and your body grows proportionately bigger. As you reach puberty, your body undergoes hormonal changes and you will see changes in your appearance. Usually, puberty starts between the ages of eight and 13 in girls and nine and 15 in boys.

In girls, the body produces the hormones progesterone and oestrogen. They are responsible for the development of breasts, hips and pubic hair.

In the case of a boy, his body starts producing a hormone called testosterone that causes changes such as the appearance of facial hair (moustache and beard), voice breaking and an increase in height.

Puberty

During puberty, your body will grow faster than at any other time in your life, except for when you were a foetus. As puberty is about to begin, your brain releases a special hormone called Gonadotropin Releasing Hormone, or GnRH. When GnRH reaches the pituitary gland (a pea-sized gland just under the brain), it releases two more hormones - Luteinizing Hormone (LH) and Follicle Stimulating Hormone (FSH). Both, boys and girls, have these hormones in their bodies.

Amazing Anatomy

Why do old people have wrinkles?

When you are young, the skin is elastic and can keep moisture inside. But with old age, the dermis starts to lose collagen and elastin (that maintains elasticity). So, the skin becomes thinner and cannot get enough moisture. The fatty layer that gives the skin a plump appearance also begins to disappear, the epidermis starts to sag, and wrinkles are formed.

Grandparents and old age

Many of us have grandparents who are old and need care. They may appear shorter and weaker every day. There are reasons behind this change. As people grow older, they generally lose some muscle and fat from their bodies as part of the natural aging process. The vertebrae may degenerate and start to collapse into one another. As they start to press closer together, a person loses a little height, becoming shorter.

As people get older, the pigment cells in their hair follicles gradually die. With fewer pigment cells in a hair follicle, the strands of hair do not contain as much melanin as before and appear more transparent in colour - like grey, silver, or white.

facts about growing up

- Beards are the fastest growing hairs on the human body. If a man never trimmed his beard, it would grow to nearly 30 feet long in his lifetime.
- The skin of a newborn baby covers an area of about 2500 cm². By the time the baby grows up, the skin would be able to cover 1.8 m² — about the size of a shower curtain!

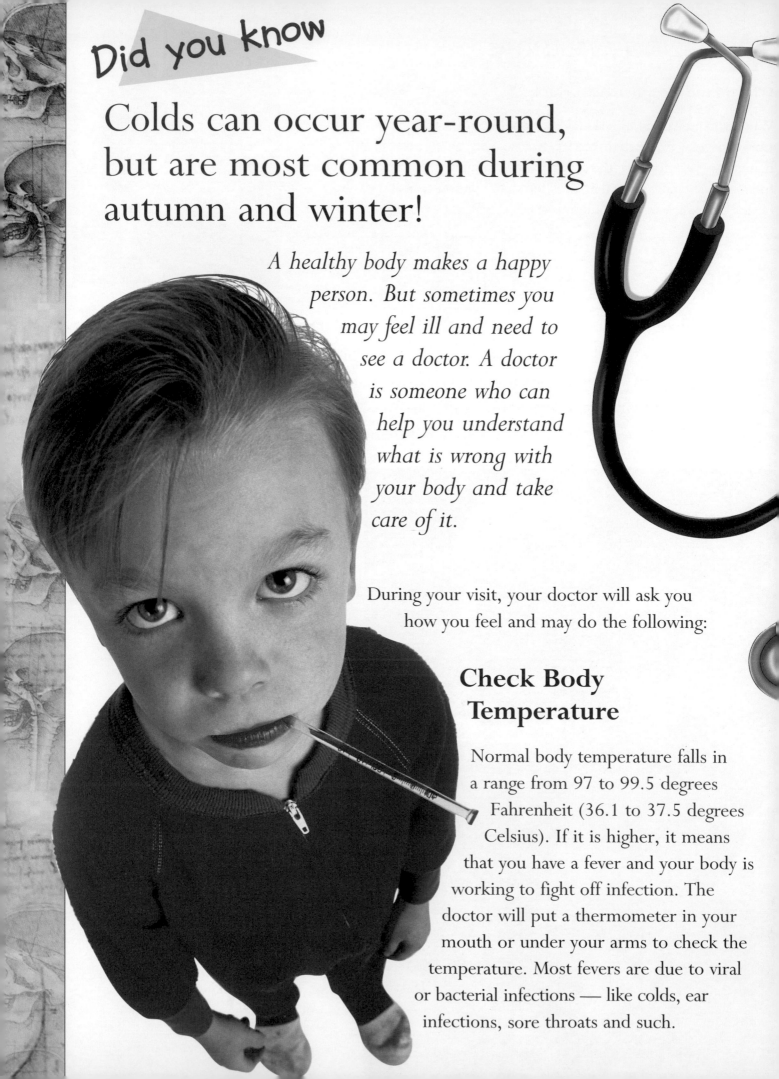

Colds can occur year-round, but are most common during autumn and winter!

A healthy body makes a happy person. But sometimes you may feel ill and need to see a doctor. A doctor is someone who can help you understand what is wrong with your body and take care of it.

During your visit, your doctor will ask you how you feel and may do the following:

Check Body Temperature

Normal body temperature falls in a range from 97 to 99.5 degrees Fahrenheit (36.1 to 37.5 degrees Celsius). If it is higher, it means that you have a fever and your body is working to fight off infection. The doctor will put a thermometer in your mouth or under your arms to check the temperature. Most fevers are due to viral or bacterial infections — like colds, ear infections, sore throats and such.

Amazing Anatomy

What is a vaccination or 'shot'?

Vaccination is the process by which your body is stimulated to produce antibodies to fight various diseases. In this method, killed or alive, but weak, micro organisms are injected into the body. When the germ enters your body, it immediately produces the required antibodies that prevent you from getting infected. The most common vaccines include ones for polio, small pox, chicken pox, measles and TB.

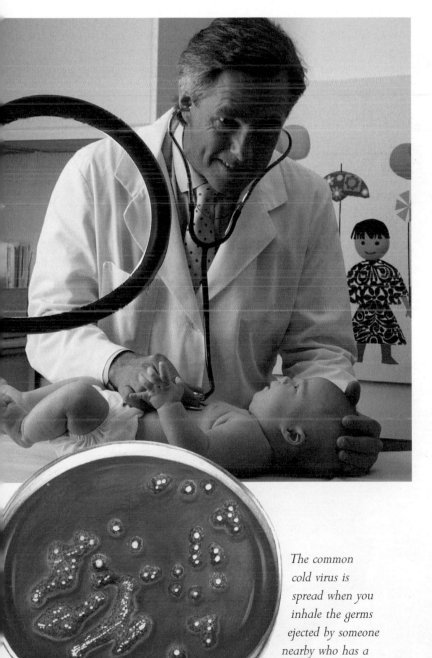

The common cold virus is spread when you inhale the germs ejected by someone nearby who has a cold, sneezes or by contaminated hand contact.

Examine Your Lungs

Your doctor will use his stethoscope and listen to your lungs. He might ask you to take several deep breaths to hear if you are inhaling and exhaling properly.

Apart from examining your lungs, the doctor also uses his stethoscope to listen to your heartbeat. He does this to ensure that your heartbeat is normal.

facts about illness

■ The first accurate drawings of the human body were done by the Flemish doctor, Andreas Vesalius (1514-1564). He used stolen corpses to conduct his studies!

■ In 1683 Anton van Leeuwenhoek, a Dutch scientist, became the first to observe bacteria under a microscope.

■ The common cold, caused by more than 200 known viruses, accounts for 22 million school days lost each year!

You need at least eight hours of sleep to remain healthy!!

In order to stay healthy, you need to take good care of your body. You need to keep your body clean and free from germs. Bathing is an essential part of your daily routine.

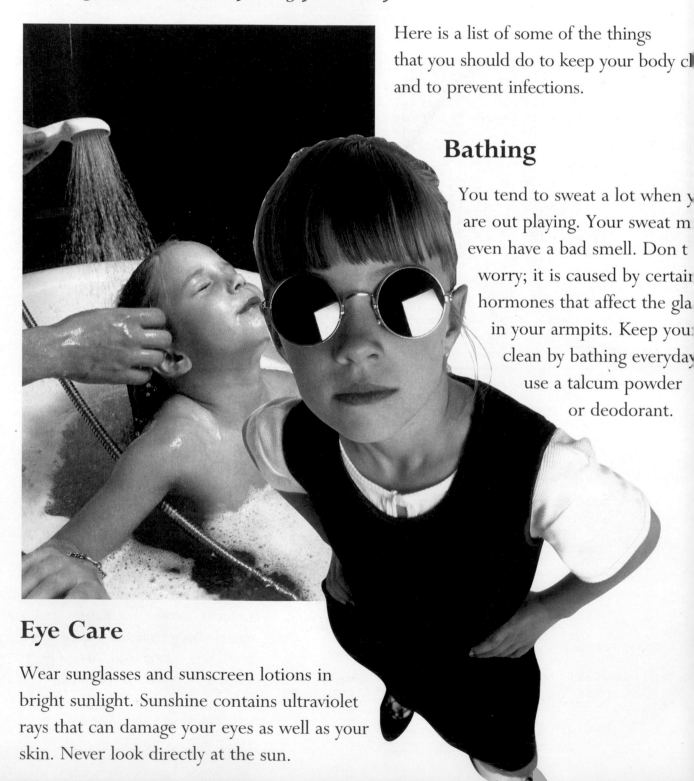

Here is a list of some of the things that you should do to keep your body cl and to prevent infections.

Bathing

You tend to sweat a lot when y are out playing. Your sweat m even have a bad smell. Don t worry; it is caused by certain hormones that affect the gla in your armpits. Keep you clean by bathing everyday use a talcum powder or deodorant.

Eye Care

Wear sunglasses and sunscreen lotions in bright sunlight. Sunshine contains ultraviolet rays that can damage your eyes as well as your skin. Never look directly at the sun.

Amazing Anatomy

How important is exercise to keep your body fit?

Regular exercise is essential to keep your body fit. Exercise keeps your heart healthy, which in turn keeps the rest of your body healthy. It also makes your body flexible and relaxes your muscles. Exercising also makes you feel happy! During exercise your body produces endorphins; chemicals that are responsible for the feeling of happiness.

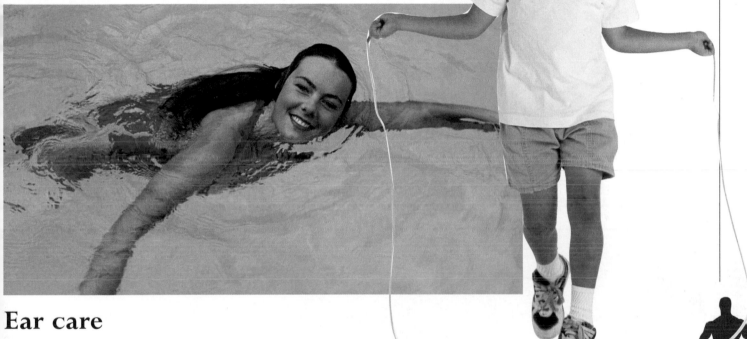

Ear care

Take care of your ears while swimming. To avoid infections like swimmer s ear, which causes swelling and pain, be sure to dry your ears thoroughly when you come out of the water. You should also rinse yourself with fresh water.

Skin care

During puberty, the oil glands in your skin tend to secrete too much oil. This, together with dead skin and bacteria, block the pores in your skin. The skin around these clogged pores can swell, and look lumpy or red. Having a fibre-rich diet, and cutting down on oily food and sweets helps a lot. Also watch out for those hamburgers and pizzas; they too should be avoided! Drink lots of water and wash your skin regularly with soap and water.

facts about bodycare

■ Lice have been around for centuries. Their remains have been discovered even on the mummies of ancient Egypt!

■ In around 1752, eyeglass designer James Ayscough introduced spectacles with double-hinged side pieces. The lenses were made of tinted glass as well as clear ones. But they were actually made to correct vision and not to shield the eyes from the sun!

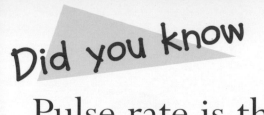

Pulse rate is the number of times a person s heart beats in one minute!

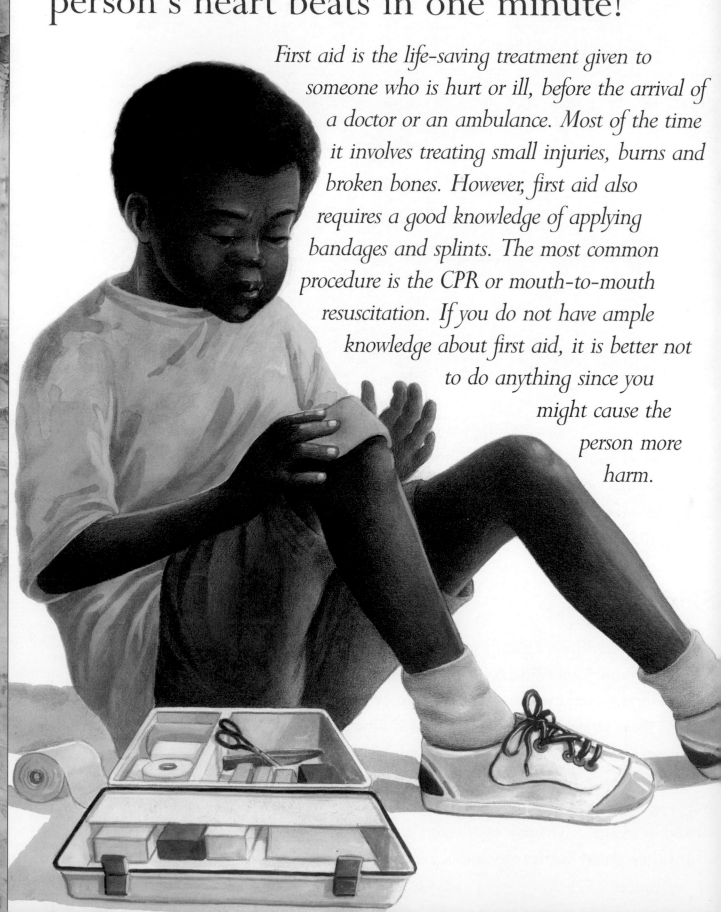

First aid is the life-saving treatment given to someone who is hurt or ill, before the arrival of a doctor or an ambulance. Most of the time it involves treating small injuries, burns and broken bones. However, first aid also requires a good knowledge of applying bandages and splints. The most common procedure is the CPR or mouth-to-mouth resuscitation. If you do not have ample knowledge about first aid, it is better not to do anything since you might cause the person more harm.

Amazing Anatomy

What is a first aid kit?

A first aid kit should contain the following: cotton for cleaning wounds, a pair of scissors, bandages, some pins to hold the bandages in place, an antiseptic cream and antiseptic lotion to clean the wounds. Always keep a first aid kit in the house.

The Red Cross symbol is usually found on first aid boxes in colours other than red.

Treating Wounds

Most minor injuries can be treated by first aid. First and foremost remove any foreign material like glass pieces from the wound. Then, clean the wound with an antiseptic like Dettol. This decreases the chances of the wound getting infected. You can follow this up with a bandage. However, if the wound is severe, one has to either raise the injured area or put pressure on the wound with a clean cloth in order to control the bleeding.

Head injury and fractures

If someone has injured his head, first check whether he is able to open his eyes or speak to you. If not then check his breathing. If he is bleeding, try and stop it by applying slight pressure with a tissue or a piece of clean cloth. Don t try to move him! Moving someone with a head injury can be dangerous. Instead, call for help.

In case the person has fractured any other bone in his body, use a splint to prevent movement of that area. A splint could be made of a piece of clean wood or magazines.

facts about first aid

■ The term FIRST AID first appeared in 1878, created by the executives of the St. John Ambulance Association as a blending of the terms "first treatment" and "National Aid".

■ The first horse-drawn ambulance was invented by Dr. John Furley in 1883!

Glossary

Antibodies: Y-shaped proteins that are produced in the human body to fight harmful organisms like bacteria and virus.

Anvil: Another name for incus, a bone found between the malleus and the stapes in the middle ear. It is called this due to its shape.

Biceps: A muscle with two heads or two points of origin. It is found in the upper arm and the back of the thigh.

Bile: A brownish-yellow fluid that is secreted by the liver. It is bitter and helps in breaking down fats in the body. It is also called gall since it is stored in the gall bladder.

Cartilage: A tough, elastic and fibrous tissue found in body parts like the joints and outer ear. The skeletons of babies are mainly made up of cartilage, which hardens to become bones as they grow up.

Cervix: The neck. The narrow outer end of the uterus is also called the cervix since it is shaped like the neck.

Coagulation: The process of clotting. A liquid substance (like blood) is transformed into a semi-solid mass.

Cochlea: A hollow space inside the inner ear that looks like a snail shell. It contains several nerve endings necessary for hearing.

Collagen: A fibrous protein found in bones, cartilage and other such connective tissue. It can be converted into gelatine by boiling.

Contaminated: To become infected or impure by contact.

Contract: To reduce in size by pulling together or shrinking.

Cytoplasm: The living contents (protoplasm) of a cell excluding the nucleus.

Deficient: Lacking in something essential (like vitamins, proteins and such).

Dermis: The sensitive layer of the skin below the epidermis. It contains blood vessels, nerve endings, sweat glands and sebaceous or oil glands.

Diaphragm: Commonly called

the midriff, it is a membrane that separates the chest from the abdomen.

Digestion: The process by which food is broken down into simpler molecules so that the body can easily absorb the nutrients.

Elastin: A protein that is very similar to collagen and is found in elastic fibres.

Epidermis: The outermost, protective layer of the skin.

Eukaryotic: Having cells whose nucleus is surrounded by a membrane.

Fertilisation: The process by which the sperm unites with the ovum (female egg) to form babies.

Glands: A group of cells or an organ that produces a secretion used in some part of the body; e.g. the sweat glands produce sweat in order to control the body temperature.

Hammer: Another name for malleus, the outermost bone in the middle ear. It is called so because it is shaped like a hammer.

Heredity: The process by which genetic characteristics or traits are carried from parents to the young ones.

Hormone: A chemical substance produced in the body that causes physiological activities like growth and metabolism.

Indole: A white crystal-like substance produced when the amino acid tryptophan is decomposed by bacteria in the intestine.

Organ: The part or structure in the human body that performs some specific function that is necessary to the well being of the body; e.g. the heart is an organ that pumps blood essential to the other parts of the body.

Organelles: A structure inside a cell (like the mitochondrion or Golgi body) that performs a specific function.

Pancreas: A long gland connected with the intestine and found behind the stomach. It is light in colour and secretes pancreatic juice during digestion.

Pancreatic juice: A colourless alkaline fluid secreted by the pancreas that helps to digest proteins, carbohydrates and fats.

Pinna: The part of the ear that is seen from outside. It is made up of cartilage.

Plasma membrane: A thin membrane that surrounds the cytoplasm of the cell. It controls the passage of substances in and out of the cell.

Platelet: An extremely tiny disc-like cell without a nucleus found in the blood plasma of human beings. It helps in blood clotting. It is also called a thrombocyte.

Puberty: The stage in which a person becomes capable of reproducing.

Reflex: An action that one has no control over; e.g. sneezing or blinking

Resuscitation: To restore consciousness or life.

Sedate: Calm, composed.

Skatole: A white crystal-like substance that is commonly used as a fixative in perfume manufacture. It has an extremely bad odour and is found naturally in faeces.

Stimulus: A substance or an agent that causes a response or reaction.

Stirrup: Also known as the stapes, it is the innermost of the three bones found in the middle ear. It is shaped like a horse s stirrup and hence the name.

Tissue: A part in the human body that is made up of cells with similar structure and functions.

Triceps: A muscle with three heads. It runs along the back of the upper arm and helps to extend the forearm.

Vomer: A thin flat bone of the nose that divides the nostrils.